P9-DWJ-772

Title page photograph:
A minke whale beginning
to surface.

# THE SEA WORLD BOOK OF
# WHALES

## BY EVE BUNTING

### PHOTOGRAPHS PROVIDED BY THE
### HUBBS MARINE RESEARCH INSTITUTE

Harcourt Brace Jovanovich, Publishers
San Diego     New York     London

# THE SEA WORLD BOOK OF WHALES

Copyright © 1980 by Sea World, Inc.
Text copyright © 1980 by Eve Bunting
Whale illustration copyright © 1987 by
Pieter Folkens
All rights reserved. No part of this
publication may be reproduced or
transmitted in any form or by any means,
electronic or mechanical, including
photocopy, recording, or any information
storage or retrieval system, without
permission in writing from the publisher.

Requests for permission to make copies of
any part of the work should be mailed to:
Permissions, Harcourt Brace Jovanovich,
Publishers, Orlando, Florida 32887.
Printed in the United States of America

**Library of Congress
Cataloging-in-Publication Data**
Bunting, Eve, 1928–
  The Sea World book of whales.
  Bibliography: p.
  Includes index.
  Summary: Describes the evolution,
physical characteristics, habits, various
species, and natural environment of the
whale.
    1. Whales—Juvenile literature. ( 1. Whales)
I. Hubbs Marine Research Institute. II. Title.
QL737.C4B868  1987    599.5    85-16409
ISBN 0-15-271948-2
ISBN 0-15-271953-9 (pbk.)
First edition

A  B   C   D   E
A  B   C   D   E   (pbk.)

Cover photo (pbk.): Female humpback
with her calf. (Flip Nicklin)

## Photography credits

Hubbs Marine Research Institute: pp. 12-left,
25-below, 68-69, 79, 81

William Evans: pp. 20-21, 33, 36-37,
43-above, 45, 61, 78

Joseph R. Jehl, Jr.: pp. 18, 19-right, 23, 26,
28-center

Stephen Leatherwood: pp. 19-left, 24, 30,
31-below, 38, 46-right, 84, 85, 86, 87, 88, 89,
90-91

Frank S. Todd: pp. 8-9, 11-above, 13, 16, 17,
31-above, 40, 43-below, 47, 53, 57, 63, 64,
65, 66, 67, 82

## Additional photography credits

Edward D. Asper, Sea World, Inc.:
pp. 32, 40

Pieter Folkens: p. 12

Robert French, Sea World, Inc.: pp. 70-left, 73

John Hall: title page, pp. 14-15

Therese Hoban: p. 11-below

Wyb Hoek: p. 44

The Kendall Whaling Museum, Sharon,
Massachusetts: pp. 58, 59, 60

Dr. Gerald Kooyman: p. 22

James Lecky: p. 28-left

Jack Lentfer: p. 46-left

David Nelson, SEACO, Inc.: p. 46

Chuck Nicklin: pp. 50, 51, 76-77

Flip Nicklin: pp. 27, 48-49

J. Olsen: p. 29-right

Hideo Omura: p. 25-above

Jerry Roberts, Sea World, Inc.: pp. 71, 72

Maurice Rumboll: p. 41

Sea World, Inc.: p. 75

Gordon Williamson, courtesy of
General Whale: pp. 54-55

Whale illustration by
Pieter Folkens with Charlotte Carlisle
pp. 34-35

# CONTENTS

Dedicated to Helen
and the Thursday Night Group.

With special thanks to
all the many employees
at Sea World in San Diego;
in particular Dr. Lanny Cornell,
Vice President, Research/
Veterinary Husbandry
and Dr. William Evans,
Director, Hubbs Marine
Research Institute.

# WHEN WHALES WALKED THE LAND
## CHAPTER ONE

**B**etween fifty and one hundred million years ago, when the world and many of its creatures were primitive, the ancestors of whales walked the land.

They were not animals that we would recognize as whales. Their heads and tails were more like those of dogs. They had bodies covered with fur, and they had four legs. They were small, too, probably no bigger than a person, although there were no humans around then for comparison.

It is likely that they waded in shallow waters near shore, sometimes swimming the edges of the sea in search of food. Gradually, their search took them farther and farther

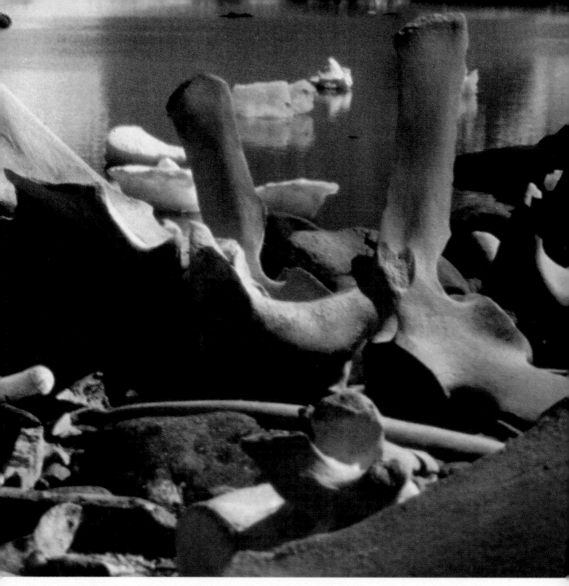

from shore.

At some point in that long-ago age some of them began to stay in the oceans, adapting slowly over millions of years to become the marine mammals that we know today as whales.

Their bodies grew more streamlined for easier swimming. Fur, which was warm when dry, but cold and heavy when wet, gradually disappeared. It was replaced by blubber (fat) under the skin. Whale blubber can be as much as two feet thick. It helps hold in heat, keeping the whale as warm as if it wore a giant overcoat.

The whales' front legs evolved into flippers, which guide the body while diving and turning. Their back legs disappeared altogether. And their tails widened to become the broad, fan-shaped flukes that whales stroke up and down to propel themselves through the water.

Their heads changed, too. The nostrils, which had been at the tip of the nose, as they are with all land animals, moved to the top of the head and became blowholes. Now whales can breathe easily while speeding along on the surface.

Those ancestors of whales probably all had teeth of some kind. Now some species of whale have them, others do not. Those with teeth — the Odontoceti (o-DON-tuh-see-tee), or toothed whales — are flesh-eaters that live by hunting fish, squid, or crustaceans (small shellfish). Those without teeth — the Mysticeti (MIS-tuh-see-tee), or baleen whales — eat plankton. Plankton is a single food source that includes a whole group of some of the smallest and most plentiful creatures in the sea. It has been called "sea soup," a good thick meat and vegetable broth that is full of the tiny animals and plants that drift in the ocean.

Baleen (bay-LEEN) whales have baleen plates growing from their upper jaws. The plates have fringed ends that hang down inside the mouth. They look like great hairy doormats, or misplaced mustaches that have grown inside instead of outside the whale's mouth. That is what the word *Mysticeti* means — "mustache whales." But baleen, unlike mustaches, is not made of hair but of horny material, similar to fingernails. The whale takes in water and plankton. Then the whale closes its mouth, pushes the water out with its tongue, and swallows the plankton that has become entangled in the baleen. Absolutely delicious ... and such large helpings!

As the centuries passed, some whales began diving deeper and still deeper in their quest for food. In the dark depths of the oceans, sight was of little value, so eyes became less important. Since fishes made noises that whales could hear, the whales' hearing became more important. Before long, some whales began making sounds of their own and listening as the echoes bounced back from underwater objects.

Since whales still had to breathe air, they improved their

Previous spread: Bleached whale bones at Port Lockroy, Antarctica.

Above: A close-up of a killer whale jaw, showing the animal's cone-shaped teeth.

Below: A close-up of the baleen of a California gray whale.

ability to hold their breath for long periods of time underwater. Heart rates slowed while diving, to use less oxygen. Muscles adapted to work without oxygen for a longer time. As a result of this, some of today's whales can dive to depths of almost a mile and stay underwater for more than an hour at a time.

We have no way of knowing how many creatures other than whales tried to return to the seas ... and failed. We do know that the bones of one marine mammal, *Desmostylus* (des-mo-STY-lus), a hippopotamus-like creature, have been found. And we know that it did not survive through the eons of time. We know also that there was once another family of ancient whales we call *Archaeoceti* (ahr-kee-o-SEE-tee). This family probably shared those first land ancestors with the Odontoceti and Mysticeti. But about forty-five million years ago, the snakelike *Archaeoceti* became extinct. Why it did is a mystery.

What we do know about past whales we learned from bits of evidence that we put together, like pieces in a puzzle. Much of this evidence came from fossils or fragments of bones found by scientists. Some of these fossils and bone fragments are over fifty million years old. The condition of the bones and the place they were found told us in which time period each whale lived. The bones themselves were pieced together so we could imagine how the animal looked.

We also learn about ancient whales by studying the whales

Opposite page: Man standing alongside a sperm whale skull. Notice the long, thin lower jaw studded with teeth.

Children kneel behind a collection of dolphin and small whale skulls.

of today.

In the developing whale fetus (unborn baby) the nostrils are still at the end of the nose. They move into place on top of the head as the fetus grows older. The fetus has some hair on its body, but most of it will have vanished by the time the whale is born. Some species keep a few whiskers for old times' sake on chins and snouts.

Small buds appear on the fetus where the hind legs once were. These, too, will have disappeared by birth. Adult whales have remnants of hip bones under their flesh. Occasionally an adult whale is found with legs that have actually grown. One was seen near Vancouver, Canada, with hind legs three feet long.

Today's whales have still other reminders of their land-dwelling past. Their flippers have finger bones similar to those in a human hand, and their several stomachs resemble those of their close relatives, the cows.

Whales are perfect examples of animals that have evolved successfully. They are now so at home with the waters in which they live that it is almost impossible to think of them in any other element. It seems more difficult to imagine a whale walking than a butterfly swimming or a horse flying. But what we have learned tells us that whales did indeed once walk the land — all those millions of years ago.

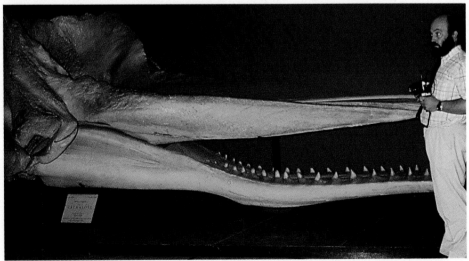

# A WHALE IS A WHAT?
## CHAPTER TWO

"A whale's not a fish though it swims in the sea
I won't eat the whale and the whale won't eat me
A whale is a what if a whale's not a fish
And you can't serve him up in a pottery dish?"

**T**he first line of the old jump-rope rhyme is true: a whale is not a fish. The second is partly true: there are no confirmed records of whales eating humans, although with a couple of species it is certainly possible. And since the boy and girl jumping rope to the rhyme are speaking English, they may never eat any part of a whale. But thousands of boys

and girls do. Whale meat is red, like beef; and, like beef, it is an excellent source of protein. In many cultures it is considered delicious. And, in many cultures where the people do not have other protein sources, it is a necessary part of a healthy diet. Some people even eat it raw. Maybe even from a pottery dish!

A whale is a what, then, if a whale's not a fish? A whale is a warm-blooded marine mammal that comes from a long and ancient line of warm-blooded mammals. That means it breathes air and suckles (nurses) its young.

Whales are called cetaceans (sa-TAY-shuns). Cetaceans are the marine mammals that have nearly hairless bodies, and paddle-shaped forelimbs.

When people speak of "the whale" they are generally thinking of one of the great whales, the gigantic creatures of the sea. But there are more than eighty different species. Some, such as dolphins and porpoises, are small whales. Others grow to be enormous.

The size of land animals is limited. Many of them have to depend on legs to carry them. A human giant would need a thigh bone so thick and heavy that it would be difficult to lift a leg to take a step. Which may be one reason that there aren't too many giants around.

Whale giants don't have legs, and their bodies are supported by water. They can grow and grow and grow. Some do. The baleen whales are the giants of whaledom, thriving on their rich diet of plankton.

The toothed whales are generally smaller than the baleen whales, with one exception — the sperm whales.

The whales' way is to go where their food supply is the most plentiful. They are migratory animals, often traveling to different oceans in different seasons, or from one shore to another of the oceans they live in. In winter, many species leave icy waters and travel to milder seas. Females often give birth in the warmer waters, where their young have a better chance of

surviving. Sometimes whole herds of whales travel together. Sometimes they move in smaller groups, called pods. Occasionally one whale will travel alone.

The whales' thick blubber not only keeps them warm but also acts as a reserve of energy, since food is often scarce on the way. Whales can and do live off this stored nourishment for months at a time.

As the whale swims, it moves its tail flukes up and down, pushing through the sea like a frogman with his feet tied together. Some swim quickly and have been observed moving up to thirty miles an hour. Others move at a more leisurely pace, going about as fast as a person would walk. Usually they travel underwater, coming up only to breathe.

The nostrils on the tops of their heads are called blowholes. Baleen whales have two; toothed whales have only one. However deep whales dive, the blowholes close and stay watertight.

When the whale surfaces, it lets out a great whoosh of breath called a spout or a blow. The warm spout makes a cloud or fog similar to the way that our breaths mist on frosty days. The spout is different in shape and size for each species. Those who know whales can tell from a distance which kind of whale is below by how high the spout goes, and whether it goes

Previous spread: A killer whale breaching in the wild.

Opposite page: A joyful trio of jumping bottlenosed dolphins.

Left: A California gray whale calf only a few hours old. Notice the wrinkled snout and the distinctive double blowholes.

straight up or at a slant.

Whales do not move in silence in their underwater world, but through the gentleness of their own sounds. Small whales make small clicks and chirps and squeaks; large whales make low moans and groans and grunts. They "whale-talk" to one another and call out to distant pods. They listen and move confidently under the guidance of their sounds. Their sounds are vibrations, much of the time too high pitched for a human to hear, but easily heard by a whale. Whales do not have vocal cords, so their sounds are not produced in the same way that other mammals produce theirs. Most scientists believe that whale sound vibrations are produced in the nasal passages.

Unlike their ocean neighbors the sharks, whales are not "swimming noses." They have lost their sense of smell. What need did they have for it? They don't breathe underwater, and when they come up to blow, their main interest is in emptying their lungs and filling them again with fresh air.

Whales see quite well, but they depend on their hearing and not on sight as they travel their long ocean journeys. Their eyes are set on the sides of their heads, not in front. So they can look not only straight ahead, but also to the right and left as they swim — or up and down, if they are swimming on their sides. To a whale,

Right: An eighty-five foot long blue whale beginning its blow. Blue whale blows can reach twenty-five feet in height.

Opposite page left: A team of researchers observes a gray whale as it blows.

Opposite page right: The characteristic V-shaped blow of the black right whale.

up and down and side to side does not mean what it does to us. Whales live in a world that is like that of a bird in the air; they are not bound by bodies that must stay on a flat plane.

Some whales travel without rest, constantly pressing on and on. Others take time out to catnap or snooze on the surface. If they sleep underwater, they instinctively come up to breathe. Perhaps each time it comes up, the whale semi-wakes, or perhaps it wakes no more than humans do when they turn in bed or drift from a dream, only to sleep again.

The sea holds dangers for sleeping whales. There may be prowling sharks, or *killer whales*, the one species of whale that preys on other cetaceans. Some of the smaller whales sleep with one eye open, perhaps allowing only half of the brain to rest at a time. Other whales that travel together may post watch whales to keep guard so that the rest may sleep in safety.

There is not an ocean of the world that is not visited at some time of year by some kind of whale. Whales have been around for millions and millions of years. Yet most humans have never seen one, and never will.

"A whale is a what?" we ask, and we quickly answer, "a warm-blooded marine mammal." But in its magnificence and mystery and beauty, a whale is much, much more.

# A SEA FULL OF WHALES
## CHAPTER THREE

**T**here are more species of toothed whales than there are of baleen whales — a case of fewer big 'uns and more little 'uns. In fact, of the more than eighty different species of whale, only ten species are baleens. Those ten species represent three baleen families:

1. the rorquals (RAWR-kwuls)
2. the right whales
3. the gray whales

The family name *rorqual* comes from two Norwegian words: *ror*, which means pleats, and *hval*, which means whale. It is easy to see how the rorquals got their name. Up to a

hundred pleats, or grooves, "stripe" their skin on the underside of their bodies. The pleats start below the chin and may run as far down as the navel. Some scientists think they are there so that the throat can balloon out and the whale can swallow a bigger mouthful of food.

Rorquals, like all baleen whales, have two blowholes. A small, sickle-shaped fin adorns their backs about two thirds of the way along the body from the snout.

The largest of the rorquals is the **blue whale**. Its name comes from its color, which is dark slate-blue flecked with white. The head is flat, the snout rounded.

The blue whale is enormous. In fact, it is the largest animal that has ever lived on earth. It is about a hundred feet long, give or take a few inches. At midpoint it is as high as a train. Its heart is the size of a small car and is so heavy that seven men would have trouble lifting it. The main artery that runs from the heart is a pipe big enough for a person to slide into.

Dinosaurs were small compared to the blue whale. Most of them could have passed through a blue whale's open jaw with room to spare. One famous photograph shows six men resting on the baleen plates inside a blue whale's mouth!

The blue whale's tall, pear-shaped spout rises twenty feet in the air and was easy for whalers to spot. Because of its body size, the blue whale was much sought after and hunted, so that now it is one of the rarest of whales.

At eighty feet in length, the **fin whale** is the second largest of the great baleen whales. It may not be the biggest, but it is still a gigantic whale: a tennis court is only seventy-eight feet long!

Slender, more graceful, and faster than the blue whale, the fin whale has been called the sea greyhound. Its color is dark gray and it has a creamy white chest, belly, and throat. The left

Previous spread: A pod of pilot whales with both adult and young whales.

Opposite page: A good look at the throat grooves of a minke whale.

Above: A large blue whale, showing its huge, volcano-like blowholes.

side of the fin whale's chin, snout, and jaw is dark gray. The right side is light gray. This may be because the fin whale often swims on its side to eat. Sea animals tend to be darker on top and lighter below for camouflage. When seen from above they blend with the darkness of the ocean. When looked at from below they are almost the same color as the light filtered through the ocean's gloom.

The **sei whale** is about sixty feet long. Its color is bluish-black, and it has a pale oval area on its chest. The sei whale's pleats are rather short and do not reach down as far as the navel. The sei (say) got its name because it appears off the coast of Norway at the same time as does the *seje*, a small fish. When swimming off the coast of Japan, the sei is called the sardine whale because then it is feeding mainly on that kind of fish.

The **piked whale**, or **minke** (ming-kee), is the smallest of

Above: A killer whale leaping. Killer whales are the top predators of the oceans.

Opposite page above:
A minke whale showing a band of white on its flipper.
This marking is present in almost all minke whales.

Opposite page below: A Bryde's whale.
These whales are identified by the two ridges
on their noses. The other rorquals have only one ridge.

the rorquals and got its name from a gunner on a whaling ship. His name was Meincke, and he accidentally shot a piked whale, thinking it was a large blue. Because of his mistake, a beautiful whale was named in his honor.

The minke is about thirty feet long, blue-gray with white underneath. Its pleats run all the way to the navel. The minke swims rather close to shore in both the Atlantic and Pacific oceans. It enters both the arctic and antarctic seas in summer and will venture into the channels and among the ice floes in search of food. Captain Scott, an antarctic explorer, described seeing it "standing upright" in ice holes with only its blowholes and snout showing.

**Bryde's whale** (BREE-dis) is rarely seen. It is a subtropical whale, and is found wherever the waters are warm. It is bluish-black with a white belly and a blue-gray throat and chin. The Bryde's whale is about forty-two feet long and is so

similar to the sei that even whale experts have trouble telling the two species apart.

The **humpback whale** can grow to be over fifty feet — as long as a school bus. It is an odd-looking whale and has been described as being "decidedly ugly." To humans, maybe . . . but not to another humpback.

Humpback whale pectoral flippers are extra long, with serrated edges, and they are so supple that the whale can arch them across its stomach or back. Its dorsal fin is rounded. When the whale dives, this fin comes up like a hump out of the water . . . hence the humpback's name. There are so many humps and bumps and lumps all over its body that the Norwegians call it "the knobbly whale."

Though its appearance may be less than beautiful, the humpback seems to have a beautiful disposition. Divers gliding in the waters alongside of these gentle giants say the humpbacks look like mammoth birds. Their long flippers seem like underwater wings on which they fly through the ocean's darkness. The humpbacks play, rolling over and over below the surface, leaping up and out into air and sunlight, crashing back in a splintering of foam. This type of jump is called breaching.

The humpbacks sing, and their songs are haunting and heartbreaking in their sweetness. In the summer of 1977, the space ships *Voyager 1* and 2 carried recordings of "The Sounds of Earth." There were greetings in fifty-five languages . . . and

Right: A humpback whale beginning a dive; its back and dorsal fin forming the hump that earned the whale its name.

Opposite page: A singing humpback whale. When singing, these whales stretch out and float almost motionless in the water.

there was the song of a humpback whale! What if whale talk turned out to be a universal language, understood by every other space creature but us? What if somewhere on some strange planet, Og is saying to Od: "Here are fifty-five recordings of gibberish . . . and just one that makes sense!"

Humpbacks are the only whales known to hunt with "bubble nets." Alone or with a partner, the humpback swims below the surface. When it spots a school of small darting fish above, it circles and blows bubbles which rise from its blowholes, trapping the fish inside the "net." Then the crowding, jumping, silvery mass on the top of the water is perfect for that great mouth and great, ballooning throat to gather in.

An ugly whale? Well, beauty is in the eye of the one who can see past the humps and the lumps and the bumps to the pure essence of humpback!

The family of right whales has smooth throats, without the pleats that are found in the throats of rorquals. With the exception of the pygmy right whale, they have no dorsal (back) fins. Right whales were given their names by old-time whalers because they were the "right" whales to catch. They were slow, easy to find, and when they died their great thick layer of blubber caused their bodies to float. Some other whales were inconsiderate enough to have carcasses that sank.

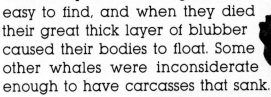

Right: Two humpback whales with open jaws, breaking the surface as they fish in their bubble net.

The **bowhead whale**, one right whale, is found only in the Arctic. It is about sixty feet long and black, with white patches on its chin. The upper jaw from which the baleen plates hang is arched into a great curving bow — the reason for its name. The baleen plates look enormous when the mouth is open. And they are. They grow up to ten feet long. That's longer than the average doorway is tall. It is hard to imagine where the whale puts them when it closes its mouth! Actually, the plates fold and fit neatly into grooves inside the whale's mouth. If taken out and laid end to end, the baleen from a bowhead's mouth would stretch for a mile.

The bowhead feeds on swarms of small floating shellfish, swimming through them with its mouth open, taking them all in. In the novel *Moby Dick*, Herman Melville says right whales

Left: An aerial photograph of a humpback bubble net.

Center: A breaching humpback whale.

make "a strange grassy cutting sound" when feeding. That was probably the sound of the baleen rattling in the water. Because of its big, strangely shaped mouth, the bowhead was thought by early whalers to be a boat eater as well as a man eater. Of course it is neither.

The **black right whale**, like the bowhead whale, is about sixty feet long and black. It has distinctive patches of rough skin on its snout, which are constantly crawling with whale lice. Whale lice are small crustaceans about the size of walnuts. They bore through the whale's skin to the blubber below, clinging with their claws to the tenderest parts.

The black right's baleen plates are not as long as those of the bowhead. This whale is often seen peering around with its head sticking out of the water, a behavior practiced by many species of whale. This is called spy hopping. The black right whale also seems to enjoy standing on its head with its tail out of the water. It is obviously a whale that doesn't miss a thing going on either above or below!

The **pygmy right whale** is only twenty feet long, truly a pygmy for a baleen whale. Black in color, it sometimes has a fine band of white running along its belly.

The pygmy right is so rarely seen that all our knowledge about it comes from studying the few that have washed ashore or stranded themselves on the coasts of New Zealand.

Why whales of any kind strand themselves is a mystery.

Below: A bowhead whale fetus (unborn baby).
Notice the "bow" shaped head
and the smooth back without a dorsal fin.

Top: A close-up of a black right's head with its jaws closed. The white spots are the areas of rough skin which are thick with whale lice.

Left: An inquisitive gray whale being rubbed by an observer.

Probably it is because of sickness or physical problems. A sick whale will usually head for land, knowing instinctively that if it loses the strength to surface for air, it will drown. Because whales are herd animals, they often follow one another onto beaches and strand together. Or whole herds may be stricken by a mass sickness like an epidemic among humans. Then, together, they head for shore.

The family of gray whales has only one species, the **California gray.**

California gray whales are about fifty feet long. They are gray, of course, but their backs have a mottled white appearance. This is because of the hundreds of barnacles that cling to the whale's body in the same way that they would to the keel of a ship. California grays leap out of the water, breaching perhaps just for fun or in an attempt to get rid of their unwelcome passengers. Like the black right whale, grays have bonnets thick with whale lice.

The California grays have the longest migration route of any group of mammals in the world, traveling approximately

A large herd of pilot whales stranded along the coast of California.

eleven thousand miles round trip each year. They move down the northwest Pacific coast to breed and mate in the protected lagoons of Mexico. Since they are not very swift swimmers, it takes them three months each way to complete the journey.

How do these whales know where they are going in their shadowy underwater world? It would seem that there are no road maps, no signposts, no service stations to pop into and ask the way. But for the grays that make the trip year after year there is a map of sorts. This is an instinct about and a knowledge of deep sea mountains and placid valleys, or rock ridges and great kelp forests that sway in the currents of the ocean, as trees bend in the wind. Scientists call this a cognitive map, a map that is carried only in the brain.

Year after year the grays return to mate or give birth, communicating with one another, wintering in the south, then returning north with their newborn calves. Some females return north pregnant with calves that will be born in southern waters a year later. They move slowly, drawn by the same invisible cords that have drawn them since their beginnings.

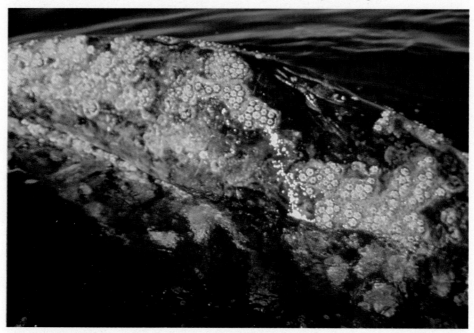

A young California gray whale covered with barnacles.

Pygmy right whale
(up to 21′)

Bowhead whale
(up to 65′)

Sei whale
(up to 69′)

Humpback whale
(up to 52′)

Narwhal
(up to 16′ without the tusk)

A Sea Full of Whales

A good look at the size, shape,
and color of seventeen different
whales.

California gray whale
(up to 46')

Black right whale
(up to 56')

Pilot whale
(up to 20')

Killer whale
(up to 31')

Minke whale
(up to 34')

Bryde's whale
(up to 46')

Giant bottlenose whale
(up to 42')

Fin whale
(up to 88')

Beluga whale
(up to 16')

Pygmy sperm whale
(up to 11')

Sperm whale
(up to 59')

Blue whale
(up to 100')

Except for the ten species of baleen whales just mentioned, all the other species of whale have teeth. These Odontoceti – toothed whales – eat either fish or mammals or both in place of the plankton that the baleen whales eat. Even though their teeth are sharp and enameled, the Odontocetes do not use them for chewing. Instead, they grip with the teeth in order to tear their prey apart. Sometimes they don't even use their teeth, but just swallow their prey whole.

While there are three families of baleen whales, there are six families of toothed whales:

1. the sperm whales
2. the Monodonts (MAH-no-donts) — single-toothed whales
3. the Delphinids (del-FIN-ids) — dolphin whales
4. the Ziphiids (zi-FEE-ids) — beaked whales
5. the Phocoenids (foh-SEE-nids) — true porpoises
6. the Platanistids (pla-ti-NIS-tids) — freshwater dolphins

These families include over seventy different species of whale. Although they are all officially classified as toothed whales, scientists further divide them into two groups: the larger toothed whales and the smaller toothed whales. These larger toothed whales, along with the baleen whales we have already mentioned, are the ones most people think of when they

think of whales.

The smaller toothed whales group is made up of all the many, many dolphins and porpoises in the sea. We do not have room to describe them all here. But we can talk about some of the larger ones.

The family of sperm whales is named after its largest member — the **sperm whale**. At almost sixty feet long, the sperm whale is also the largest of the Odontocetes. It is the most "whale shaped" whale — the one that is seen painted on signs that hang above places like The Whaler's Inn or Moby Dick's. It is the whale-shaped whale most first graders will draw when asked to draw a whale.

One-third of the sperm's body is made up of its huge, almost square-shaped head. Its blowhole is slightly off-center and leans to the left. Because of the blowhole's position, the sperm whale's spout also leans to the left and is easy to identify.

The sperm whale's teeth are set in the lower jaw only. But what teeth they are! There are twenty to thirty of them — gigantic, cone-shaped, ivory, and occasionally thick with barnacles. The upper jaw has neat sockets. When the whale closes its mouth, the teeth fit into the sockets as cleverly as pegs into holes.

Sperm whale teeth were prized by whalers. The whalers

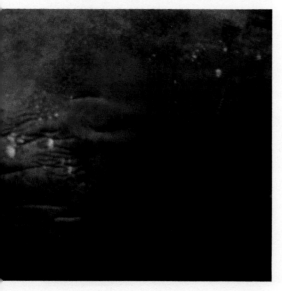

A dead sperm whale at a whaling station. The teeth in the sperm whale's lower jaw fit neatly into sockets in the upper jaw when the jaw is closed. The head is covered with scars made by the sharp beaks of giant squid. Sometimes pieces of the beaks are left imbedded in sperm whales' heads.

pulled them from a dead whale's mouth and scratched designs on them with sailmakers' needles. The scratches were then darkened with soot from the ship's lamps. Often the designs were whaling scenes, or pictures of great, full-blown sailing ships, or wonderful animals that the whalers had seen on their travels to strange, far-off lands or forlorn seas. The art was and still is called scrimshaw. Making scrimshaw helped sailors while away the time on long, becalmed, dreary days at sea. Scrimshaw was easy to sell back in port . . . a genuine souvenir of the romantic whaling life . . . a real ivory, sperm whale's tooth. And today it is even more valuable.

The sperm whale has a rounded dorsal (back) fin, which is the first in a series of bumps that run along its back, close to the tail flukes. The whale's color is dark gray, sometimes shading to silver-white underneath.

The name *sperm* comes from a waxy substance that the whale has in its head. Old-time whalers thought that this substance contained the whale's sperm, its seeds of life, and so they named it *spermaceti*. Although this spermaceti does not carry any sperm, it is an exceptional substance. It was used for lubricating fine machinery and watches and was especially good for making fine candles. Spermaceti candles are practically smokeless and burn brightly. The altars of many churches

Scrimshaw and sculptured articles made from sperm whale teeth.

glowed with the strong steady flames of candles made from spermaceti.

Sperm whales live, feed, and breed mainly in tropical seas although male sperm whales are seen in the arctic and antarctic. They are deep divers, searching out giant squid in the dark depths of the oceans. These enormous squid, the sperm whale's favorite food, have eight arms and two slender tentacles covered with suction discs. Hidden under the writhing arms is the parrot-like beak with which the squid tears its prey to shreds. Sperm whales swallow the squid whole, but the beaks are hard for them to digest.

Sometimes the beaks lodge in the whale's stomach, and a greasy substance forms around them. Now and then the whale will cough up a lump. Boulders of it, up to nine hundred pounds in weight, have been found floating in the sea. More often it stays lodged in the stomach. This substance is called ambergris (AM-ber-grees). It was, and still is, used in the making of expensive perfume, and at one time was very valuable. Its value has decreased in modern days because chemical substitutes have been developed. Old-time whalers received extra pay when they found a glob of ambergris. It was an added and welcome bonus to their whaling money.

Sperm whales can handle most things without trouble. A shark over ten feet long was found in the stomach of one, along with the skin of a seal, and eight feet of fishing line with six hooks attached to it. Bucketfuls of sand have been found inside sperms. One sperm had a glove in its stomach. This led to fears that it had swallowed a man, and the glove was all that was left. More likely it was just a floating glove.

Big and bulky, a sleeping sperm whale can sometimes get in the way of a boat. Or, more truthfully, the boat gets in the way of the sperm. In World War II, a United States destroyer hit a sperm whale during the night. The jolt was so great that the men on board thought they had been torpedoed and took to the lifeboats. Next morning, the sperm whale was found impaled on the destroyer's bow.

The sperm whale has two small relatives. One is the **dwarf sperm whale**, and the other is the **pygmy sperm**. They both grow only to about thirteen feet. Their heads are rounder than that of the sperm, and more in proportion to the rest of the body. Both of these whales have teeth shaped like those of the sperm, but smaller. The only difference between them is in the number of teeth they have – the pygmy sperm from 24 to 36 and the dwarf sperm from 16 to 22.

The **giant bottlenose**, or **Baird's beaked whale**, is the largest whale in the Ziphiids family. It can grow to over thirty feet in length and, among the toothed whales, is second in size only to the sperm.

It is not nearly so toothy as the sperm whale. The giant bottlenose has only two pairs of teeth in the lower jaw, and none in the upper jaw. Its color is slate-gray or brown with white blotches, and it has a beak that sticks out from under its bulging forehead. There is spermaceti oil in that forehead, too, though not as much as in the head of a sperm whale.

The giant bottlenose eats mainly squid, and lots of them. Thousands of horny squid beaks were found in one bottlenose stomach — all that was left of many squid dinners.

The **killer whale**, the largest member of the Delphinid family, is streamlined and powerful. Its colors are black above and white below with a white patch above each eye and a saddle of gray behind its tall dorsal fin. A full-grown male may

Above: Baby sperm whale brought to Sea World in Orlando, Florida for rehabilitation.
Opposite page: A killer whale with a full grown sea lion in its mouth.

be thirty feet long and weigh fifteen to twenty thousand pounds. Females are smaller, growing up to twenty feet and weighing eight thousand pounds. The male's tall back fin may be six feet high. It looks like the sail of a boat as it slices through the water.

The killer whale has large cone-shaped teeth in both jaws. They are set into the upper and lower jaws in an alternate pattern, so that when the jaws close, the teeth fit together, interlocking like a trap.

Killer whales are always on the hunt for food, moving and working in packs of up to twenty or thirty. Though their main diet is fish, they will also attack seals, birds, and other whales. They will even put sharks on their menu. They are voracious eaters; the remains of fourteen seals were found in the stomach of one killer whale. The bones were probably what was left of several meals and not simply one giant feast.

There has been only one human death attributed to a killer whale, and that one was not proven. On occasion, though, humans have felt threatened. When Captain Scott was on his expedition to the South Pole in 1911, one of his men reported that killer whales tried to knock him off an ice floe and into the sea. The killer whales probably mistook him for a seal. Or they may have been bumping the ice to make channels through which they could breathe.

Killer whales have no natural enemies. They are the top predators of the oceans. Killer whales act as wolves of the sea, weeding out the weaker creatures as wolves on land take out the sick and diseased deer or antelope. Whatever is fleeing from a killer whale doesn't have much of a chance. These whales have been observed swimming at speeds of over twenty miles an hour.

Willie Willoya, an Eskimo whale hunter, tells of seeing a pod of killer whales chasing a small group of blue whales. The blues could tell they were going to be caught, and, according to the Eskimo's story, "had a conversation." Then one old one swam back and sacrificed himself, allowing the others to get away. The Eskimo story is a nice one but scientists don't accept it.

The **pilot whale**, like the killer whale, is a member of the Delphinid family. The pilot whale got its name from fishermen who knew that these whales fed on herring and other schooling fishes. They believed the whales led, or "piloted," them to where these fish were.

Pilot whales are one of the most commonly seen species of whale. They swim in all the oceans of the world.

The pilot's body is dark gray or black. Its belly has a distinctive light spot on it. The light spot is always in the shape of an anchor. The pilot whale grows to a maximum length of about twenty feet. The head and forehead are very round. In fact, because of its shape, the pilot whale's forehead is called a melon.

Pilot whales commonly swim in large schools and sometimes strand themselves in masses on beaches. Efforts to refloat them have never been successful.

The **narwhal** is a member of the Monodont (single-toothed whale) family. It is found only in northern waters near the edges of ice, and is twelve to sixteen feet long and blue-gray in color. Its head is rounded with a hint of beak, and it has no dorsal fin.

The narwhal has only two teeth; one on each side of the upper jaw. In females both these teeth stay under the gums, but in the male, one of them — usually the left one — grows and grows. It ends up being a wonderful tusk about eight feet long.

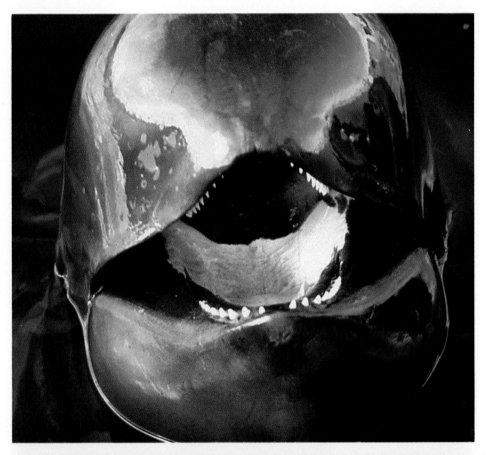

Above: A close look at the head of a pilot whale. Because of their distinctive "pot" shaped heads, pilot whales are often called "potheads."

Below: A pilot whale swimming off San Clemente Island, California.

Narwhals are sometimes washed up on beaches or trapped in bays by ice. Early whalers would bring the tusks back from arctic voyages. No landlubber could imagine what kind of creature had grown such a tusk, and so a magical beast was invented to solve the mystery. The beast was the unicorn, a type of horse which was of course "seldom seen." Its horn was supposed to have grown from its forehead. And the legend spread. The horns, when ground into powder, were said to cure diseases. In the days of Queen Elizabeth I of England, a tusk sold for what would today be five million dollars.

What use the narwhal has for its tusk is still not known. Perhaps it uses it as a fencing weapon. In 1957, an Eskimo found a narwhal that had a three-foot long tusk of its own and five inches of another buried in its upper jaw. Some scientists believe the tusk may be used as a tool to pierce holes in thin ice when the narwhal is frozen in and needs to breathe. Another

Right: An aerial view of a pod of narwhals. The narwhals' long ivory tusks make them easy to recognize.

Opposite page: A good look at the creamy white beluga whale.

more likely theory is that the tusk is both an ornament that the male uses to attract the female, and a weapon that he uses against other males.

Whatever its use, the narwhal tusk is something quite unusual. Not many creatures can boast of a single splendid ivory tusk, long and spiraled and beautiful.

Like narwhals, **belugas**, or **white whales**, are members of the Monodont family. In fact, these two species are the only species in that family. The name *beluga* (buh-LOO-guh) comes from the Russian word "*belukha*," meaning *white*. These whales grow to a length of about fourteen feet and prefer icy waters. Because of the high-pitched chirping sounds belugas make, sailors called them sea canaries. These whales have about ten teeth on each side of both upper and lower jaws and feed on fish, squid, and small crustaceans.

Like narwhals, belugas are often trapped under the ice

when the seas freeze. The small, open areas that are left become filled with jostling whale bodies searching for air. Then they are easy prey for hunters.

A Soviet scientist saw a polar bear lying near a hole in the ice. Twenty or twenty-five belugas were trapped in the hole. Each time one of the whales surfaced to breathe, the bear smacked it on the head and dragged it out onto the ice. Thirteen belugas had already been pulled out on the ice beside the bear.

Belugas move freely between salt water and fresh. Often they swim into rivers to feed, particularly when salmon are spawning and there's a chance of a tasty meal. In order to save valuable salmon supplies, a biologist with the Alaska Fish and Game Department and a biologist from the U.S. Navy developed a kind of beluga repeller. They tape-recorded the sounds of killer whales and played them back on underwater tape recorders. The recorders are placed in the water at river mouths and are working well to scare belugas away from

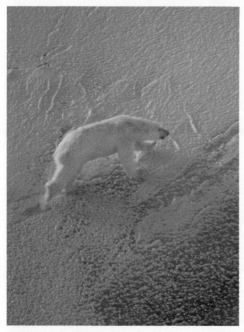 

Above left: An Alaskan polar bear, one of the main predators of beluga whales.

Above right: Polar bear tracks surrounding a breathing hole in the ice.

valuable salmon.

There are reports, though, that the belugas are getting braver. Hunger and greed are overcoming fear, and the belugas are discovering that killer whale sounds don't necessarily mean that killer whales are really there. How confusing. And how confusing it must be in the oceans when they disregard killer whales' sounds and find that this time the noises are backed up by those great black-and-white bodies – with those bottomless appetites for beluga whales.

The neck bones of belugas are not fused together as in other whales. They can turn their heads from side to side — and should, when they hear those killer whale sounds coming fast in their direction!

There are more whales and still more: the **false killer whale**, the **ginkgo**, **Hubbs beaked whale**, and all the many dolphins and porpoises in the oceans, too many to mention here. There are suborders and families and subfamilies that together make a whole, limitless sea full of whales!

A false killer whale in a graceful leap.

# A FAMILY AFFAIR
## AFFAIR
### CHAPTER FOUR

If a male whale is showing off his swimming skill, waving his flippers, or flipping his flukes, he may be only playing around. If he is leaping into the air, and leaping again, he may be simply frisky. But if this is happening in the right place at the right time, the chances are that he's showing off to attract a female whale.

There are no known mixed-breed whales in the wild. As far as we know humpback whales do not mate with blues, or grays with narwhals. The reason for this may lie with those waving flippers and flipping flukes. Some scientists think that each special type of showing off charms only that special type of whale.

Other scientists believe there are no mixed breeds because like only attracts like. It may be that simple.

Many species of whales are monogamous. They stay with one mate at a time. Killer whales and sperm whales, though, are known to be polygamous; that is, the males have many mates at the same time. A bull sperm may have a harem of fifteen or more females. And sometimes he has to fight to keep them away from other male sperm whales who want to take the harem over.

Fighting male sperm whales charge toward each other, their great mouths agape. They will ram one another, or butt heads, or tear each other's flesh. What a strenuous life, keeping all those mates together and battling off all that competition.

Whales are seldom seen while mating. They usually live their lives far from the eyes of humans. And when we do chance to come upon a pair, it is difficult to see exactly what is going on. Usually the observers are on the deck of a ship, peering down into water that is white with churned-up spray, and bright with reflected sunlight. Or they may be looking through binoculars across distances of ocean. Those who have had a brief glimpse of the process say that whales mate belly to belly as they swim. Or "stand" belly to belly straight up and down in the water.

Although most whales are large creatures, the male's sperm are as small as those of men. The fertilized egg clings to the wall of the mother's womb. Over the months, the unborn baby grows there. It will be a year to sixteen months, depending on the species, before it is ready to be born.

The poet sounds despairing when he says:
    "I think that I shall never see
    a baby blue about to be!"
And that is true. He probably never will. The birth of a great blue whale has never been seen by man, and may never

be. Blues are very rare, and swim the far reaches of the oceans. But enough whale births have been recorded to let us know basically what happens.

Baby whales generally back out into the world: they are born tail first. There have been exceptions to the rule, though. Scientists have observed at least one head-first birth in both gray whales and beluga whales. And their observations have led them to believe that a head-first birth occurs occasionally in other whale species as well.

The normal tail-first entry into the world is one of nature's safeguards. Soon after its head is outside its mother's body, the baby must breathe. And to breathe, it must get to the surface. Animals born head first in the water risk the possibility of drowning.

The umbilical cord separates almost automatically as the baby leaves the mother's body. Then the mother nudges the baby toward the surface, anxious that it take its first life-giving breath.

Scientists from the University of California watched a gray whale give birth off Baja California. They saw the head emerge first. Seconds later the calf was on the surface. It had already taken three breaths when the mother came up underneath

Previous spread: A female humpback whale and her calf.

Opposite page: Two right whales rolling and touching one another during the mating ritual.

Left: An aerial look at a black right whale with her calf.

and lifted it clear of the water. Mothers never can resist butting in!

A Soviet scientist on a research ship saw a baby sperm whale immediately after birth. A large female was in a group of about forty whales. One quarter of her body was raised above the surface as she "trod water." Her calf, about twelve feet long, was still attached to her by the umbilical cord. The calf's tail was limp and curled as though it had not yet had time to have a good stretch after being inside its mother for so many months. Two other females underwater seemed to be supporting the baby.

This behavior may not be unusual. Other whales often appear to "help" during a birth. This has led scientists to believe that there may be midwife-like helpers in the more social whale societies — those societies where many whales travel together. Humpbacks, sperm whales, belugas, and killer whales are very social whales. These species have often been observed helping one another in times of stress or sickness.

The baby whale is born open-eyed and ready for action. Almost as soon as it has taken its first breath it begins to swim. To nurse, since it has rather rigid lips, it opens its mouth wide and swims close to the underside of its mother's body. The mother has a nipple hidden in a slit in her skin. She contracts the strong muscles in her milk glands to help squirt the good, rich milk into the calf's waiting mouth.

Whale mothers often nurse their calves while lying on their sides with one flipper in the air.

Whale milk is thick, more like heavy cream than milk. In his book *Lost Leviathan*, Dr. F. D. Ommanney says: "Once we collected some and had it for breakfast with our porridge (oatmeal). It was sweet and very fatty and on the whole spoiled a good plate of porridge."

It may have spoiled the doctor's porridge, but it does what it's supposed to do for baby whales. Those big, strong babies need strong, hearty food. At birth, baby sperm whales are about thirteen feet long. A newborn blue whale weighs about two tons and is twenty to twenty-five feet long. That's as long as a two-story house is tall. During its first weeks of life, a baby sperm whale gains ten pounds every single hour. It drinks about a hundred and fifty gallons of milk a day. Try to imagine three hundred half-gallon cartons of milk lined up side by side. It's a good thing mother doesn't have to have it delivered!

Baby whales nurse for five to eight months, depending upon the species. They stay close to mom for food, protection, companionship, and education. For even though baby whales are born with certain abilities, like the abilities to see, swim, breathe, and hold their breath, they need to mimic their mothers in order to learn to use their skills.

It's a full-time job being a whale mother — so much of a job that most whale mothers couldn't manage more than one baby at a time, without one of the babies or themselves starving. So nature takes care of the situation. Twin babies are very rarely, if ever, born, and mothers only give birth once every two years. So there's plenty of time to give to that one big baby — to prepare it well to survive in the harsh sea world that will be its home.

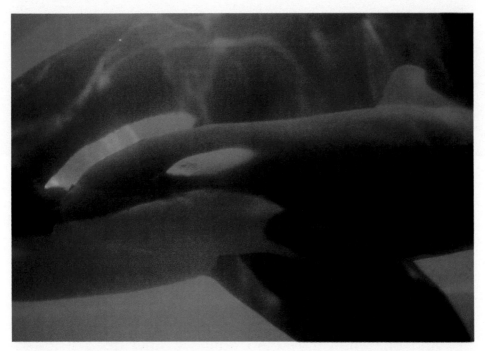

A killer whale with her newborn calf.

# FIN OUT
# IN THE WATER!
## CHAPTER FIVE

It is not hard to imagine how it was, way back in time, when primitive people saw their first dead beached whale. How they must have pondered this giant creature. What was it? A great fish? A monster from the depths? They would circle it perhaps. Prod it. They would alert the other villagers, who would come to where the tide washed around the great dead body. Fear would give way to curiosity. They'd move closer. One, more daring, would clamber up the slope of the back. Feet would pound on that huge mound of solid flesh.

And oh, when they tried eating it how good it tasted! Cooked or uncooked! And how much there was of it!

"When one whale is caught it makes seven villages prosperous," says the Japanese proverb. It is no wonder that ancient peoples stood on ancient shores and prayed for the gift of another whale. And no wonder that they tired of waiting and took to the waters in their little boats in search of a whale to catch. Fish were such small mouthfuls. But a whale ... now there was a prize worth having.

Primitive whaling must have been terrifying for those men in the little boats. They were exploring the then new world of the sea, much the same way as our modern-day astronauts explore the new world of outer space. Primitive whalers had only spears or knives for weapons. Catching a whale was a dangerous test of skill and the measure of a man's worth. It is likely that in early days, many, many boats were lost each time whaling parties set out. But the rewards were worth the risks, for those wonderful whales gave not only meat and blubber for people's bellies, but oil for their lamps. The whale's massive bones could brace the walls of huts. Their skins and sinews were turned into tools for survival. Human minds looked for safer and easier ways to catch whales, and human minds found them.

When North American Indians spotted a single whale, they would surround it in their small boats and attempt to kill it with arrows, then tow it to shore and share the feast.

Eskimos in the Canadian arctic used sticks to catch beluga whales. They knew the belugas swam into rivers to feed when the tide was rising. When the belugas were upriver, the Eskimos stuck long, thin sticks in the sand at the river's mouth. As the tide ebbed, the belugas began their swim back to the sea. The sound of the water strumming against the sticks frightened them. They milled around and often stranded themselves on sandbanks.

Norwegians used noise to drive fin and sei whales into narrow fiords. They would go out in small boats, circle behind a pod of whales, and bang stones or rocks together to herd the whales shoreward. When the pod entered the fiord, they closed off the opening with nets so that the whales were trapped.

People were whaling out of necessity. Whales were food and food was life. There wasn't a bit of the whale that went to waste.

The first European people to carry on organized whaling were the Basques in the twelfth century. The Basques (basks), who live on the coasts of France and Spain, built lookout towers on shore. When the watchmen saw whales spouting, they rushed to light fires of wet straw on the beach. The people of the town saw the smoke and gathered on the shore, ready to take to the boats.

By this time whaling had developed more sophisticated weapons. The harpoon, an arrowhead prong with a line attached, held the caught whale close to the boat so it could be killed.

Only certain species of whales could be taken. Some swam too far from shore for a small boat to follow. Some were too fast and could outswim any oarsman. Some sank when they died: such a whale could never be hauled up from the ocean's bottom.

Large, slow whales that swam close to shore were perfect to chase. They could be easily spotted, overtaken, and towed to land. The bowhead and all the right whales were the "right" whales to catch. Smaller whales, such as the white, or the pilot, or the narwhal, were good too. But they were smaller and gave less oil and blubber. Others, such as the large blue, the fins, and seis were too difficult to catch and bring in.

In the thirteenth and fourteenth centuries the Basques began building bigger ships that could go farther out to sea. And other nations were looking at whaling and wanting a part of it.

Previous spread: A whale-catcher boat with two whales in tow.

Above: A whale bone arch made of jaw bones. This arch was erected in 1933 to commemorate 100 years of British rule in the Falkland Islands.

By the sixteenth century the Dutch and the English had whaling fleets too. Often they hired the rugged Basques for their crews because "they were excellent men for the killing of the whale."

When the first settlers arrived in America, the American Indians were already taking whales for food and oil. The new settlers were soon whaling too. By 1775, as many as sixty ships were sailing out of New Bedford, Massachusetts, in search of whales to satisfy the growing demand for whale products in the Colonies and on the Continent. By 1846, the American whaling fleet had grown to seven hundred twenty-nine ships, and whale products were fueling the machines of the fledgling industrial revolution.

Yankee whaling ships were heavy, wooden vessels with

Above: A Dutch whaling scene entitled *Whaling in the Northern Sea Ice*, painted by Abram Salm around 1700. The Dutch flag is clearly seen.

Opposite page: A painting of a Yankee whaling ship passing the Groton monument in Groton, Connecticut, by E. Baker entitled *Whaleship George*. Notice the wooden hull with big square sails.

big, square sails. The crews' quarters below decks were dark, cramped, smelly, and damp. The men slept on straw mattresses that they called donkeys' breakfasts. Their beds were bunks, built one above the other in narrow rows. The food was poor, the flour often black with bugs, and crawling with worms. The crew lived in a stench of grease and blubber.

During the daylight hours, one lookout was always at the top of the mainmast, and one was at the top of the foremast. There was a bounty of ten pounds of free tobacco for whichever one spotted the whales first.

Each ship carried four or five smaller boats. Each of the smaller boats held a harpooner, a steersman, and at least four oarsmen. At the lookout's shout of "Thar she blows!" the boats were lowered.

The rowers were not allowed to look over their shoulders as they pulled toward the whales. If they had seen that the mammoth bulks were very close, they might have been tempted to pull the other way! It was the harpooner's job to make the strike on the whale. It was a dangerous job requiring great skill and strength.

Frank Bullen, a whaler himself, tells of the terrors of being in that small boat, rowing or sailing toward the monstrous whale. Sometimes the crew became so frightened that they "required a not too gentle application of the tiller to their heads in order to keep them quiet."

And well they might be frightened. Although cases of a whaler being swallowed by a whale were rare, it was an extremely dangerous business from start to finish. Many whalers were drowned when their small boats were upset or "stove in" by the whale's powerful flukes. Sometimes the struck whale would drag the boat for short distances as fast as twenty miles per hour. This was called a Nantucket sleigh ride. Other times the whales pulled the small boats so far from the sailing ship

that they were lost.

If they caught their quarry and got it back to the whaling ship without incident, there were more dangers to be faced. The massive body was tied to the side of the ship, its weight dragging the vessel down to port or starboard.

The flensing or "cutting up" was done from a platform slung over the side of the sailing ship. The great chunks of blubber had to be hauled aboard by hand, and dragged up on that slippery, tilting deck. The blubber was boiled out in large pots on the deck, so the oil could be stored in casks for the journey home. There was always the terror of a fire on that wooden boat, miles from anywhere in the middle of the ocean. And these perils had to be faced day after day, week after week, month after month on long whaling voyages.

Why would a man become a whaler? For some there was no choice. A man might be knocked on the head and wake hours later with a bump on his skull, his money gone, at sea on a whaling ship. *Shanghaied,* it was called, and there were gangs that made quick money shanghaiing men onto ships in

Opposite page: An engraving by Martens of a painting entitled *Pêche de la Baleine* by Garneray. The work was done in the 1800s and shows a right whale about to be taken.

Left: The flensed throat area of a fin whale. Notice the throat grooves and the alternating layers of meat and blubber.

need of crewmen. The fee was the new sailor's first month's wages in advance – and no questions asked.

But there were good ships too, with good reputations, good food, and a good feeling on board. For these ships there were waiting lists of crewmen because there was money to be made on a successful whaling trip. And whale products were in demand.

Humans had always known the value of whale oil. Rich and fine, its flame burned bright in lamp and candle. But in the eighteenth century, with the start of the industrial revolution in Europe, the need for fine oil increased. There was no petroleum then, no kerosene. Whale oil was the only oil available to run the machinery — the spinning Jenny, the power loom, fine watches and clocks — and whale oil was essential. Men set out on year-long to three-year-long voyages to fill the holds of their ships with the precious oil to keep the wheels turning.

And there was more to a whale catch than just the oil. There was meat. There was bone meal from the bones. There was vitamin A in the liver and insulin in the pancreas. And there was baleen.

Baleen could be used in the making of brushes and for the hoops in ladies' crinolines, for whip handles, and indeed for anything that required springiness and strength. After its use for fashion was discovered, the baleen from one whale could pay the expenses of a whole whaling voyage. Everything else was profit.

Small wonder that the demand for whales grew. And as it grew so did the demand for more ships. When there were more ships there was more whaling and a demand for still more ships, one industry supporting the other one.

Until the middle of the nineteenth century, though, the only whales being taken were still the slow ones that floated when they were dead — the right kinds of whales. The other kinds were left alone. Then steamships were introduced, and the faster whales that traveled far from land could be chased and caught. These new steam-powered whaling vessels were larger and sturdier than the earlier sailing vessels and therefore safer for the whalers.

An abandoned catcher boat at Gritvyken, South Georgia, in the south Atlantic.

The invention of the harpoon gun in 1846 by a Norwegian whaling captain, Svend Foyn, made whaling easier still. The new harpoon carried an explosive and could be shot into a whale from a cannon mounted on the deck of the steamship. Now only one correctly placed shot was necessary to kill a whale. From this point on, there were no whales that were too big to be taken. But the bodies of some of the biggest still sank. In this respect they were still not the "right whales" to catch.

It was Captain Foyn who thought of a solution to this problem too. Compressed air was pumped into the dead whales to make them float. Then they could be held or towed to shore. By the end of the nineteenth century there were in reality no "wrong whales" in the oceans.

It is still basically Captain Foyn's type of harpoon and his method of inflating whales that are used today. Modern whaling ships have radar, sonar, and some have helicopters. The catcher boats that operate from the mother factory ship follow the whale herds as they move around the oceans. The gunner stands high on the bow and aims for the whale's back, between the shoulders.

When it is dead, or "fin out" in the water, the whale is inflated, flagged, and has a radio transmitter attached to it. The

Opposite page: A harpoon gun.

Below: Two types of exploding harpoon tips.

Far left: A scientist holding marking harpoons. These harpoons, fitted with radar reflectors, were used to mark the dead whales so that they could be spotted and picked up by the factory ship.

catcher boat is then free to chase and harpoon another whale.

The factory ship that picks up the dead whales is often almost as big as an aircraft carrier. It processes the whales on board. Then huge tankers transport the oil, and refrigerator ships return the meat to port.

Modern whaling has come a long way from that first primitive human on that far-off primitive beach. It has come so far that it is nearing an end. The factory ships are now inactive because the larger, more profitable whale populations have been depleted. In addition, the International Whaling Commission regulations say that the only whales that can be taken are minkes, and they can be taken only in the Antarctic. At thirty feet long, the minke used to be considered too small to be worth the bother, so there are more of them left than blues or seis or fins.

Whaling existed when we needed it. The old-time whalers braved danger and death to bring us precious oil and whale products. They sailed uncharted, unfriendly waters, and they drew maps and opened the seas for those who followed.

Much of our information on whales comes from examining the whale carcasses brought in to whaling stations. We learned about whale growth and age, and what whales ate, and how

Right: An abandoned shore whaling station at Gritvyken, South Georgia. This station was shut down in 1965.

they reproduced. Many of the men who sailed on whaling ships were not only interested in catching whales, they were interested in whales. They were there — and they kept logs. Their records helped us to understand the habits of whales, their patterns of living.

Today, whaling is a dying industry. Many species are commercially extinct. Commercially extinct does not mean that the species is gone forever. It means there are so few of them left that it is no longer profitable for the big factory ships to go searching for them. It costs more in fuel and manpower than the price they get for the scattering of whales they find. In the end this will do more to save the whales than any regulations.

Countries that still whale will use the boats they have while they are still usable. Many of these boats are already old and rusting. When they are ready for the scrap heap, they will not be replaced.

As whaling ends, the whales have their chance to grow in numbers again as they did when whaling was shut down during World Wars I and II. Nature has a way of healing. When a population decreases, the birthrate increases. So let it be for whales. Good luck to them. May their seas be fair and filled with peace.

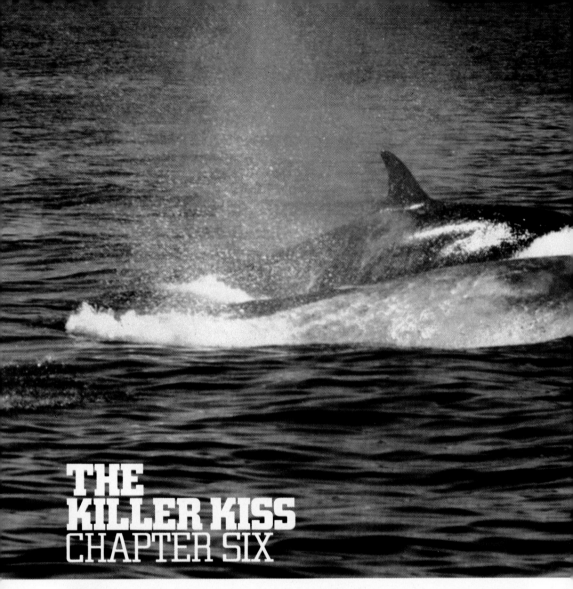

# THE KILLER KISS
## CHAPTER SIX

**K**iller whales are killers. They are the top predators of the oceans, afraid of no other creature, not even the mighty shark. Working in packs, they feed on seals and dolphins. They will move in on other whales twenty times their size. So terrifying are they that when gray whales sense the presence of killers, they have been reported to go into shock and lie, belly up, waiting to die.

There are no records of killer whales attacking humans, but there are no records to say they have not done so. There is nothing to say they would not do so. What would make a human so special that a killer whale would not want to eat him,

given the chance? They have been called "the most efficient killers in the seas," not only because of their businesslike teeth and jaws, but because of their intelligence.

In 1979 scientists on the research vessel *Sea World* from Hubbs Marine Research Institute, San Diego, saw and photographed a pod of about thirty killer whales attacking a sixty-foot blue whale. The killers moved in like a pack of wolves. Some swam on either side of the blue to keep it from escaping. Some went ahead and some stayed behind to cut it off fore and aft. One group swam below the whale to make sure it couldn't dive to get away. Another group swam on top in what looked like a smart move to keep the blue from coming up for breath.

Then they attacked. They began pulling off great chunks of flesh. The dorsal fin was stripped away. The tail flukes were chewed off. One gaping hole dug in the blue's side was more than six feet square. The attack on the living whale went on for about five hours before the killers quit and swam away. Perhaps they had had enough to eat. Or perhaps they just tired of the sport. The blue whale, barely alive, swam on.

Bruce Stephens, Director of Animal Behavior at Sea World, San Diego, says killer whales are easy to work with because they have no fear. "Dolphins are timid," he says. "Killer whales will come right up to you."

Sure they will. But why? What do they have in mind? They are super big creatures with super big teeth. Do they forget that they once used to kill? Remember, their reputation in the wild isn't exactly peaches and cream.

Bruce laughs, "These whales aren't hungry." Bruce believes the ones in the wild behave as they do because they need to hunt to keep themselves fed. They are carnivores, meat eaters. Everything in the sea feeds on everything else. Killer whales just happen to have rather large appetites. And of course

they hunt in packs. They are intelligent, and that's the intelligent way to hunt.

Bruce Stephens doesn't necessarily believe all the gruesome stories about killer whales, either. "Just about anyone could handle one," he says. Presumably he means killer whales that are as well fed as those at Sea World. But of course, not just anyone does handle the killer whales, even at Sea World. The trainers who train the whales are first trained themselves.

Dave Butcher is Director of Animal Behavior at Sea World in Florida. "When a new trainer begins work," he says, "I tell him to sit at the edge of the pool and watch the animals. Just watch and watch and watch. Sometimes the new trainer will say to me, 'You mean this is my job? Just sitting here?' and I'll tell him, 'Quite right. Just sit here and get a suntan. And watch.'"

To Dave, observing animals is the most important thing a

Previous spread: Killer whales attacking a large blue whale.
Opposite page: Shamu and a trainer nose to nose in the water.
Left: Shamu performing a leaping behavior at Sea World.

killer whale trainer or any animal trainer can do. In time, the trainer can begin to sense the personalities under those elegant black-and-white skins. He can see what one whale likes, or dislikes. One may prefer being alone a lot of the time. One may like company. Sometimes it's a little of each, the way it is with people. Like people too, killer whales have moods, and their moods show . . . if you know what to look for.

The trainers train "the Sea World way." It is training with feeling and grows from understanding and respect — and affection. No animal is ever kept hungry so that it will do anything for food. The animals are well fed at all times

Bruce Stephens jokes, "The whales at Sea World have triple chins! Have you ever seen a whale with a triple chin?" Not quite. But all the Sea World whales, and seals and walrus, are pleasantly plump.

When a killer whale does a behavior correctly, it is "reinforced" . . . rewarded. The reward *could* be a tasty tidbit. But it is just as likely to be a belly scratch, or a flipper rub. It will be something the trainers have learned that the whale really enjoys. Under the trainer's guidance the whale will stretch, close

its eyes, and look blissfully happy. There is a sense that if it could purr, it would. The whale might be given a favorite toy during playtime. The killer whales at Sea World love to pull the big, water-filled ball underwater. And the ring that floats on the surface is great to nose bump or haul around.

The killer whale kiss, which is so incredible to see, is, in Bruce Stephens's words, "one of the easiest behaviors to teach."

First Shamu was taught to stick his tongue out. This wasn't hard because killer whales love to have their tongues and gums rubbed. Then he was taught to put his tongue on the trainer's hand. To do this, the trainer simply laid his hand against Shamu's tongue a few times. Each time tongue and hand touched, Shamu was reinforced.

Pretty soon, when the trainer moved his hand Shamu would come up, find it, and put his tongue against it. Then the trainer began bending over and laying his hand against his own cheek. Shamu would come up and "lick" the hand. The trainer began taking his hand away and rewarding Shamu each time he put tongue to face. "What a good boy. That's just what I wanted." One trainer then went to the other side of the pool and

Above: Shamu having his tongue rubbed.

Opposite page: Shamu (foreground) and Namu with their trainers at Sea World.

another one would give Shamu the signal to go over and "kiss." Like every other behavior, it was done step by step, with encouragement and love.

The pretty girl's name is Beverly. She's from Iowa and she's vacationing in California. Today she stands by the side of the pool waiting to be "kissed" by Shamu. She's giggling, self-conscious, nervous. What things are going through her mind? She knows about killer whales. She knows about those big, sharp teeth. She's heard killer whales are the super killers, more fierce than sharks. So now she's going to stand here and let one of these ... these killers ... come out of the water right by her head? She's got to be crazy! But this is Sea World. I mean, they wouldn't let her do this if it was dangerous, would they?

"Ladies and gentlemen ... Shamu," the trainer's voice booms.

Beverly hears a great underwater surge, but she doesn't dare turn her head because she's been told to stand this way, sideways, and if she turns. . . .

There's an enormous whoosh of water and she hears the "ooh's" and the gasps from the audience. The huge body is a black-and-white blur that she senses more than sees, and the big, lollopy tongue is like a washcloth against her face. Somewhere a whistle sounds. There's a smooth flash as Shamu drops back into the pool below, and everyone's applauding. Beverly breathes again and laughs and shakes water from herself like a puppy dog. She uses the towel the attendant gives her to wipe her face.

"That was terrific," she tells him. "I can't believe it. I never even saw a killer whale before, and now I've been kissed by one. Wow!"

"Now I ask you, Beverly, is this whale a killer ... or a lover?" the microphone voice booms.

"A lover," Beverly says shakily, her words almost drowned in the applause from the crowd.

Killer whales are killers. They are the top predators of the oceans, afraid of no other creature, not even the mighty shark. But this beautiful, smiling, friendly killer whale, Shamu, couldn't be a real *killer*, could he? What a silly question!

The killer whale kiss.

# GETTING TO KNOW THEM
## CHAPTER SEVEN

**T**here was a time when people thought whales were fish that defended themselves by blasting water from their blowholes. Drawings showed torrents of water hurtling down from holes in whales' heads onto ships, sweeping hapless whalers to their deaths. Now we know better.

Once people thought that some whales were magical, eating "nothing but darkness and the rain that falls in the sea." Now we know better.

Slowly we are getting to know whales. And our knowledge is coming step by careful step.

In the days of Aristotle, scientists nicked the tails of small

whales and released them. If the whales were caught again later, the scientists would know how far and from where these whales had traveled. It was a primitive form of "tagging," something that today's scientists still do in a more complicated way.

Today, scientists on research boats tag whales by shooting harmless stainless steel tubes into their blubber. A number is etched on each tube and a reward given for its return. This technique began in 1923, and because of it scientists have had a better understanding of the movements of whales ... where they go in different seasons.

Because many years may pass between the placing of the tube and its return, tagging is also a way of finding out how long whales live. Unfortunately, the returned information tells only how long a particular whale lived from the time it was tagged until the time it was captured, not how long it could have lived. So tagging, while it increases our overall knowledge of whales, still leaves a lot of questions.

Up until the middle of the twentieth century, research on whales was limited, for the most part, to tagging information and to what we could see of the animal when it came up to breathe. Below, in the ocean's darkness, it was lost to us. Much of what we knew about whales physically came from the examination of dead ones. And since creatures in death are not really like creatures in life, we had many mistaken ideas about how whales looked and how they functioned.

The turning point in whale study came in the 1950s, when the first small whales could be successfully kept in oceanariums. Then we could not only marvel at the grace and beauty of the living animals, but also learn from them.

We learned that whales empty their lungs and fill them more completely with each breath than do humans. We learned that they make better use of the oxygen in the air than we do. These are two of the reasons why they can dive so deep and stay underwater for such long periods of time.

Early experiments had been done to test the theory that

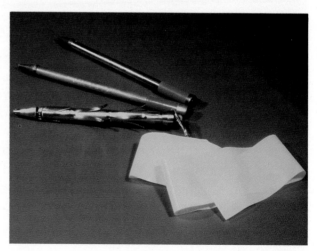

Previous spread: Divers photographing humpback whales.

Right: Research discovery tags. These tags are harmlessly shot into whales' blubber for research purposes.

Opposite page: A beluga whale blindfolded with latex eyecups.

small whales can navigate by the use of sonar. A dolphin was blindfolded by having latex cups attached painlessly over its eyes with suction. The blindfolded dolphin was placed in a tank filled with hanging metal pipes. It swam unerringly through the maze. A killer whale and a pilot whale that had been trained to bring back rings tossed into the water could find them as easily when blindfolded. Later experiments showed that blindfolded dolphins could select live fish from dead fish in their pools. They also could tell the difference between objects made of iron and objects made of aluminum.

Before the opportunity came to study whales in captivity, "ageing" (finding the age of) whales was done by "the educated guess." People who had spent much of their lifetimes in the study of whales in the wild could estimate whale age by whale size. Or, in the dead whale, by the condition and development of some of the internal organs. Estimates were that killer whales probably lived about twenty-five years, grays about fifty years, sperms about seventy-five years, and that blues might make it to the ripe old age of ninety-five. But the question of age was the biggest hole we had in modern whale knowledge. It still is. We are working on it, though, and we are making progess.

Baleen whales have cone-shaped wax plugs up to three feet long in their ears. When these plugs are cut lengthwise, rings can be seen in the wax like the rings in the trunk of a tree.

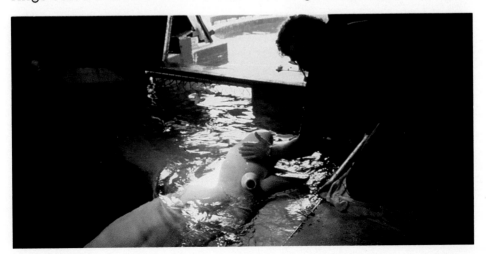

Perhaps each ring represents one year of the whale's life. Or each ring may represent two years . . . or only six months. Unfortunately, no one knows for sure.

Whales' teeth have similar rings. Was each one of these rings one year of life? Again, no one knew. Now, working with toothed whales in oceanariums, we may find out. Programs have been started at Sea World, San Diego, to give some of the small toothed whales a type of medication that leaves flourescent rings on their teeth. When a different tooth is examined at regular intervals, the rings can be counted under a microscope. Scientists then know how much time has passed "between rings." There is hope that soon ageing of these kinds of whales can be done accurately.

Familiarity with whales in captivity led to new inventions that taught us more about whales in the wild, too. Telemetric devices were developed, as were radio packs that could be attached to whales. Use of these gave us new understanding of what whales in the oceans did at night as well as what they did during the day. And what they did underwater as well as what they did on the surface

The advantages of having live whales available for study became more and more obvious. But keeping captive whales healthy is no easy job. The water in their pools must be pure and always at the right temperature. Food quality has to be constantly checked.

Whales suffer from many diseases, just as humans do. The whales have to be carefully watched. In the wild the whale's way is to hide its sickness and try to look normal. A healthy-looking whale has a better chance of survival in a sea filled with predators searching for an easy meal. This natural instinct for cover-up makes it difficult for veterinarians working with whales. They have to be detectives as well as doctors, always on the lookout for any signs of trouble.

Whales in oceanariums are immunized to prevent disease before it strikes. They have regular blood and urine tests. Some are trained to hold up a fluke or a flipper on signal so a blood sample can be easily taken. A sort of "roll up your sleeve, this isn't going to hurt" in whale talk!

Left: Researchers
working with captive
bottlenosed dolphins.

Below: The veterinary
laboratory at Sea
World.

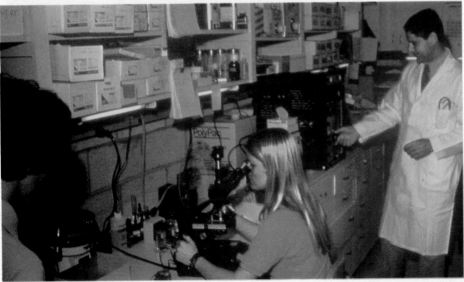

Individual veterinarians make individual discoveries. Dr. Lanny Cornell, veterinarian at Sea World, San Diego, says that a sick whale taken from the water for treatment will sometimes stop breathing. He found that it can be encouraged to start again by having water poured over its head. The whale feels the water crossing its blowholes and thinks that somehow it has surfaced. Then the instincts of years take over, and it will breathe again.

Caring for comparatively small, toothed whales is difficult enough. To try to collect and keep a gigantic baleen whale seemed impossible. Scientists had resigned themselves to knowing these big whales only fleetingly in the wild ... or passively in death.

And then there was Gigi.

The year was 1971. Sea World, San Diego, got permission from the Mexican and United States governments to go to Scammon's Lagoon in Baja California and bring back a baby

A newborn California gray whale.
From its position in the water we know that this
baby is riding on its mother's back.

gray whale. There, where the grays came yearly to mate and give birth, a gray calf would be easily found. A great deal could be learned about baleen whales if this mission could be successfully accomplished.

The scientists were sure that the baby would make it safely from its sea world to theirs. Sure? Yes, up to a point. It is difficult to be totally sure about the reaction of any wild thing brought among people.

The plans were made to collect and maintain a calf for a year or two, and then return it to the sea. The expedition under the direction of Dr. David Kinney, Sea World veterinarian, set off.

On March 11 they dropped anchor in Scammon's Lagoon. There was one large boat, the *Margaret F.*, and a smaller one, the *Martha Jane*. It would be the *Martha Jane* that would try to capture the calf.

For seventeen hours the crew tried for one calf after another. But the gray mothers were too clever. They kept their babies moving and out of harm's way.

The crew turned off the motors and let the *Martha Jane* drift. In the calm silence a mother and calf surfaced close to the boat.

The *Martha Jane*'s engines roared into busy life as mother and calf dived for safety. When they came up to breathe a man in the bow dropped a lasso around the calf's neck. Quickly he eased it down the length of the body and tightened it around the tail. Another noose was dropped over the head. They had their calf! Carefully they began towing it toward shore.

The mother did not give up her baby without a struggle. But soon the calf had been pulled into shallow water where the mother could not follow.

Those in the boat were sad to part mother and baby. But they reminded themselves of why they were doing this. And they reminded themselves too of the love and care that waited for Gigi back in San Diego. The calf was a female and that was what her name would be ... Gigi.

All night long she was kept wet and comfortable on the beach. When morning came she was maneuvered into a sling and floated on rubber tires and empty oil drums to the *Mar-*

*garet F.* There her sling was winched on board, and Gigi was transferred into a huge pool for the three-day trip to San Diego.

The pool had to be huge. The two-month-old calf was eighteen feet long. That's longer than a car. She weighed forty-three hundred pounds, which made her as heavy as twenty men.

From the minute of capture, Gigi showed no fear. She seemed to watch those who cared for her with more interest than alarm. When she was placed in the fifty-five thousand gallon pool that waited in Sea World, she began to get better acquainted with humans. They came into the water with her. They put a rubber hose in her mouth and squirted food through it. The food was a mixture of cod liver oil, vitamins, yeast, ground-up squid, bonita, corn oil, and water — something like a huge, nutritious shake! For Gigi, although it wasn't exactly mother's milk, it was good.

She grew and grew and grew.

Before long she had to be moved to a larger pool that held a hundred and ten thousand gallons of water. And she was given a playmate for company. Her new pal was a female bottlenose porpoise called Speedy. Speedy became so possessive of Gigi that she bothered Gigi's handlers, clapping her jaws at them to leave Gigi alone! Speedy ate solid food, and soon Gigi was following suit, scooping up frozen squid from her tank.

And still she grew. And again she needed more space. Her

next new pool held a million gallons of water. Would she never stop growing? This baby was gaining thirty-six pounds a day — one and one-half pounds per hour!

She had a mind of her own. She liked some of her trainers, and she didn't like others. Those not in her favor soon knew about it. Gigi would go into a tantrum when one of them appeared, rearing up and lashing out with her flukes. A woman called Sue and a man named Bud were her chosen ones. She even let Sue ride around the pool on her back.

It was Bud who taught her hand signals, and before long Gigi could respond to "tap commands." Two taps on her head meant she should open her mouth. Three taps meant "No! Stop it, Gigi! Be quiet!"

The attendants and Gigi would "talk" together, Gigi rumbling a low-pitched rumble, grunting, bonging, beeping. Scientists recorded the sounds. Why and how did Gigi make these sounds? They weren't sure. And why wasn't she making the clicking noises that grays were thought to make in the wild? Were clicks used only as communications with other grays? Did Gigi not use them now, knowing she was the only gray whale around? The scientists weren't sure about this either.

Researchers monitored Gigi's breathing, examined her blood, listened to her heartbeat. They measured the thickness of her skin, the depth of her fat and muscle. Many of these tests had never been done before on a large baleen whale.

Opposite page: Gigi and Speedy with two researchers.

Left: Sue riding on Gigi's back.

The public came to see her. She was enormous! She seemed so gentle. Was she really smiling at them through the glass, or did all whales look like that? "She's so cute!" people said. Cute? At twenty-four feet long and weighing over ten thousand pounds, that was a lot of cuteness. Gigi was the largest captive animal in all the world. And still she grew.

Toward the end of the first year it was obvious that the million gallon pool would soon be too small. What to do next? It was unthinkable to hold an animal in an area that might become uncomfortable. Gigi was twenty-seven feet long now and weighed seven tons. She would be even bigger.

In July, Dr. Bill Evans of the Naval Undersea Center was called in. He and the other whale experts talked over the situation. In February the gray whales that had spent the winter in Scammon's Lagoon would be starting north again. It was decided that Gigi would be released in the waters where they swam, in the hope that she would join them and become part of a group of migrating whales. It was a difficult decision, laced with danger and controversy. Would Gigi adapt to life in the wild, or would she fall easy prey to the sea's carnivores? Would her natural instincts help her find food that wasn't on the floor of a

pool? Those who worked with Gigi felt she would be all right. She was intelligent. She was healthy, strong, and well fed. Certainly she was in good condition for the new adventure. And it was a chance that had to be taken.

There was only one last thing Gigi could do for science and that science could do for her. She would carry a radio tracking device so that her movements could be traced. If she had problems, humans would know. They would know where she was. Perhaps she could be helped.

A radio pack was attached to her back. The stitches that held it to her skin were put in under local anesthetic so Gigi felt no pain. The pack was designed so that with time, and the help of salt water, it would corrode and fall off. A mark was painlessly freeze-branded on her, too. It would be Gigi's own mark and would identify her for the rest of her life.

Speedy the porpoise was taken from the pool and returned to the other porpoises. The friends had to be parted now. Speedy would not be going.

Gigi was placed in a huge sling and taken by truck to the Navy barge which was to carry her out to sea. Bud and Sue and Gigi's other friends stayed with her to the end, calling out their

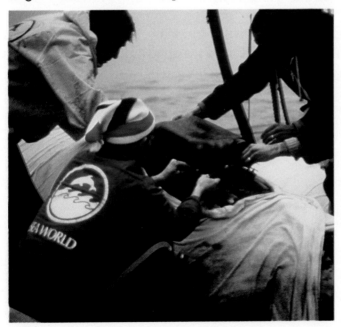

Opposite page: Gigi with an admirer.

Left: Scientists applying Gigi's radio pack.

good-byes as she was lowered into the ocean.

It was a heart-stopping moment. What would she do?

Gigi swam around the barge and around again, unsure of herself in this great pool without walls. Then she lowered her head and dove, her flukes fanning first the air, and then the water, disappearing as she dove.

A hydrophone had been lowered into the sea, and those in the boat listened for Gigi's first free sounds. And then they came — long trains of clicks, more than a thousand of them, the kind scientists had listened for in vain in the pool. Who was she calling to? What did she say?

For the next few days the trackers reported on her movements. She joined up for a time with a group of traveling grays. Then she was alone again. Then it was reported that she was with another group.

People kept seeing her. She surfaced close to a wharf where a fisherman stood. She scared swimmers and surfers by coming up and spouting so close that they could almost touch her. Gigi was a sociable whale. Hadn't she been with people almost from the beginning?

For about six weeks the backpack sent in its messages. But the transmissions were getting fainter. One observer said that kelp was snarled in Gigi's radio antenna. Possibly that was interfering with the signals. The last message came two months

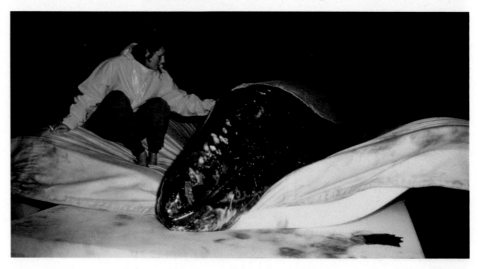

after Gigi had been released. She was not heard from again.

But she was seen the next year ... and the next, and two years after that. Gigi just kept popping up. A group of children from a Los Angeles elementary school out on a whale watching boat had a whale pop up close to them. Their teacher was sure that it was Gigi. She took pictures and contacted Dr. Evans in San Diego. Back at school the children made drawings of the playful whale that had surfaced so close to them. Dr. Evans was invited to talk to the children and see the art display. And there were the whale pictures, complete with Gigi's own special freeze marking on the back!

There will probably not be another Gigi experiment. No other baby gray whale will ever be captured, kept, and released. The Gigi experience was unique. It was a lesson in trust, and in new understanding for all those who came to know Gigi.

And what was it for Gigi? Does she still remember a time when her world had walls? When a woman rode on her back and a man taught her with hands of love? Does she remember Speedy? Does she ponder the strangeness of it all with other whales as they swim together, arching their dark bodies through the endless oceans? Maybe they whaletalk: "I think they are finally getting to know us. They are beginning to understand that whales are special, that if we disappear from this earth we can never be replaced."

Opposite page: Gigi and Sue.

Far left: Gigi in her sling on the way back to sea.

Left: Sue's tearful good-bye to Gigi.

The whales surface, blowing out the great whoosh of their whale breaths, filling their lungs with the sparkle of morning. They are immense. Their backs are sliding mountains. Their flukes spread the ocean as they dive, and there remains only empty sea and a rainbow that hangs, like a promise, over the water.

# INDEX

Midwife, 52
Migration, 16, 77
  of gray whale, 32
Milk, 52-53
Minke, 24-25, 34-35
Moby Dick, 29
Monodonts, 36
  See also Beluga; Narwhal
Monogamy, 49
Mothers, 50-53
Mouth, 10, 23, 29
  of baby, 52
  of bowhead, 29
Mysticeti, 10
  See also Baleen whales

# N

Narwhal, 34-35, 42-45
Nasal passages, 18
Navigation, 79
Neck, 47
Norwegians, 56
Nostrils, 10, 13, 17
  See also Blowholes
Nursing, 52-53

# O

Odontoceti, 10, 36
  See also Toothed whales
Oil, whale, 62
Oxygen, 12

# P

Perfume, 39
Personality
  of gray whale (Gigi), 85
  of humpback, 26
  of killer whale, 72
Phocoenids, 36
Piked whale, 24
Pilot whale, 34-35, 42, 79
Plankton, 10
Platanistids, 36
Play, 26, 73
Pleats, 21, 24, 25
Pods, 17
Polar bear, 46
Polygamy, 49
Porpoises, 16, 36, 84
Predators, 19, 68
Products, from whales, 38, 39, 62
Pygmy right whale, 30, 34-35
Pygmy sperm whale, 34-35, 39-40

# R

Radio packs, 80, 87
Reproduction, 48-53
Research methods, 76-88
Rest, 19
Rewards. See Trainers and training
Right whales, 20, 28-31
  "right one" for whalers, 28, 57, 65
Rings, age, 79, 80
Rorquals, 20-28

# S

Salmon, 46
Scammon's Lagoon, 82-83, 86
Scott, Captain, 25, 41
Scrimshaw, 38
Sea World, Florida, 71
Sea World, San Diego, 69-74, 80, 82-89
Sei whale, 24, 34-35
Senses
  hearing, 10, 18
  sight, 10, 18
  smell, 18
Shamu, 73-74
Shanghaiing, 61
Shape, of whale, 37
Shark, 39, 41, 68
Sickness, 32, 52, 80, 82
Sight, 10, 18
Singing, 26
Single-toothed whales, 36
  See also Beluga; Narwhal
Size, 16, 34-35
  of Gigi, 84, 86
  See also under individual whales
Sleeping, 19
Smaller toothed whales, 36, 37
Smell, sense of, 18
Social behavior, 19, 52
Sonar, 10, 79
Songs, 26
Sounds, 10, 18
  of beluga, 45
  of feeding, 30
  of gray whale, 85
  of humpback, 26
  of killer whale, as beluga
    repellant, 46
  See also Sonar
Species, 16, 17, 20, 36
Speed, swimming, 17

# BIBLIOGRAPHY

## Books and Journals

*Alaska Whales and Whaling*, Vol. 5. Alaska Geographic Society.

Coerr, Eleanor, and Evans, Dr. William E. *Gigi: A Baby Whale Borrowed for Science and Returned to the Sea.* New York: G. P. Putnam's Sons, 1980.

Coffey, David J. *Dolphins, Whales, and Porpoises: An Encyclopedia of Sea Mammals.* New York: Macmillan, 1977.

Coleman, Hewett, Berres, Briscoe. *Whale Hunt.* Field Educational Publications, Inc., 1967.

Daugherty, Anita E. *Marine Mammals of California.* Sacramento: California Department of Fish and Game, Second Revision, 1972.

Graham, Ada, and Frank. *Whale Watch.* New York: Delcacorte Press, 1978.

Harrison, R. J., ed. *Functional Anatomy of Marine Mammals,* Vol. 2. New York: Academic Press, 1974.

Journal of the Fisheries Research Board of Canada, Vol. 32, No. 7. 1975.

Kirk, Ruth, and Daugherty, Richard D. *Hunters of the Whale: An Adventure in Northwest Coast Archaeology.* New York: Morrow Junior Books, William Morrow, 1974.

Mackintosh, Neil Alison. *The Stocks of Whales.* Coward and Gerrish, Ltd., 1965.

Miller, Tom. *The World of the California Gray Whale.* Santa Ana, Calif.: Baja Trail Publications, Inc., 1975.

Mörzer Bruyns, Captain W. F. J. *Field Guide of Whales and Dolphins.* Amsterdam: Tor, 1971.

Mowat, Farley. *A Whale for the Killing.* Boston: Little, Brown, 1972.

Ommanney, F. D. *Lost Leviathan.* New York: Dodd, Mead & Co., 1971.

Ridgway, Sam H., ed. *Mammals of the Sea.* Springfield, Ill.: Charles C. Thomas, 1972.

Scammon, Charles M. *The Marine Mammals of the Northwestern Coast of North America.* New York: Dover Publications, Inc., 1968.

Scheffer, Victor B. *A Natural History of Marine Mammals.* New York: Charles Scribner's Sons, 1976.

Scheffer, Victor B. *The Year of the Whale.* New York: Charles Scribner's Sons, 1969.

Small, George L. *The Blue Whale.* New York: Columbia University Press, 1971.

Tryckare, Tre, ed., *The Whale.* New York: Simon and Schuster, 1968.

Walker, Theodore J. *Whale Primer.* San Diego, Calif.: Cabrillo Historical Association, 1975.

*The Whale Manual.* Friends of the Earth, Ltd., 1978.

## Papers

Telfer, Nancy, M.D., Cornell, Lanny H., D.V.M., Prescott, John. "Do Dolphins Drink Water?" Reprinted from the Journal of the American Veterinary Medical Assn. Vol. 157, No. 5.

Cornell, Lanny H., D.V.M. "Marine Mammal Medicine."

Wood, F. G., and Evans, W. E. "Adaptiveness and Ecology of Echolocation in Toothed Whales." Hubbs-Sea World Research Institute, San Diego, California.